So You Want to Be a Caterer

(On a Budget)

Diane Svenwol Olson

*To Richard,
Thanks again!
With Love,
Diane Svenwol Olson*

VANTAGE PRESS
New York

The recipes contained herein have not been tested by the publisher.

Cover design by Susan Thomas

FIRST EDITION

All rights reserved, including the right of
reproduction in whole or in part in any form.

Copyright © 2008 by Diane Svenwol Olson

Published by Vantage Press, Inc.
419 Park Ave. South, New York, NY 10016

Manufactured in the United States of America
ISBN: 978-0-533-15956-7

Library of Congress Catalog Card No.: 2007909364

0 9 8 7 6 5 4 3 2 1

To my mom and dad, Gerald and Grace Svenwol

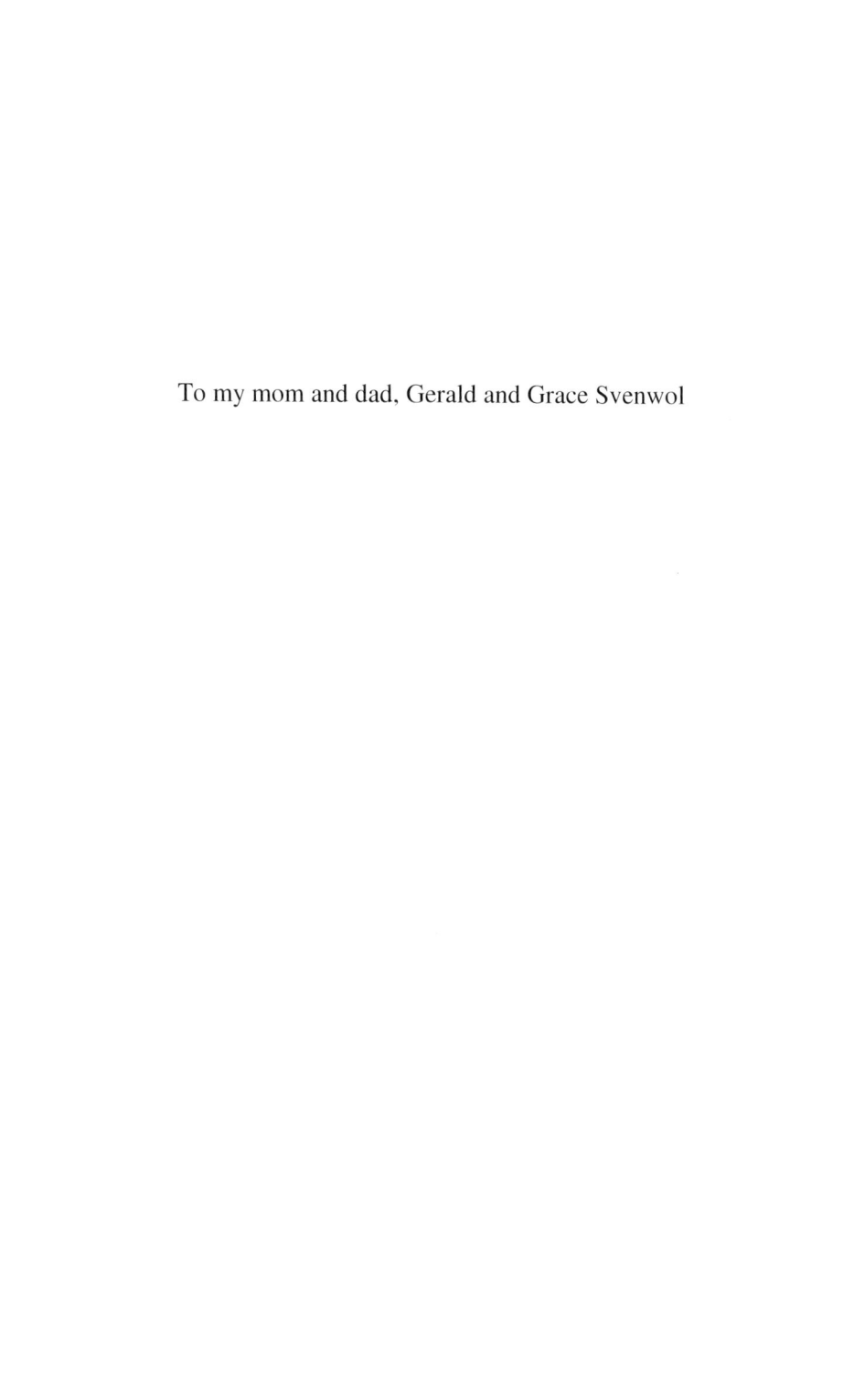

Contents

Acknowledgments   vii

Gratuity Included	1
Catering	4
Learning the Hard Way	8
Soldier Field	10
Recipes	13
The Church Lady	20
Happenings	22
The Baker Hotel	24
The Table	27

Acknowledgments

I would like to thank all those who worked with me and encouraged me with the book.

Al, Vicki, and Lisa Will, and Lora Kaiser, for being there to set up all the tables and understanding when I changed my mind as to where they were to be placed.

Deb Hansen, for all your help in working at Mr. K's and getting this book ready for printing.

Bob Krarup, Jr., for all the carting and carrying food up and down stairs to wherever we were serving.

Grace Dorpals. There is no way I can repay you for all your help. You are one in a million and you taught me a lot. Hope you get a laugh, or two, when reading this book. There are many more tales to be told and when we get together there is a lot to talk about.

Dave Christensen and Jeff Pepper, for doing what had to be done without being told. I'm sure your work ethics have stayed with you as you became adults. You were in high school when you worked for me.

Elaine Damler and Elsie Lee, who both died too young but shared with me their baking skills, recipes and ideas, and encouraged me when I needed it.

Barbara, Cindy and Chris Lundgren, who worked together to make our presentations appealing to the eye. Even their parents got into the act occasionally and it is fun to hear from them and know what beautiful people they turned out to be.

The Nelson boys, Bob and Richard. They now own their own jewelry store in Crystal Lake, Illinois. They were both very good at whatever I asked them to do. They are still that way with all their customers.

The purveyors that worked with me: White Cross Bakery, Paul's Deli, A to Z Rentals, M&M Meats, Crichton's, and Piggly Wiggly.

And for my friend, Mary Link, who kept after me to write, write, and write some more. It was at her suggestion that I joined the creative writing group at the Muskego Library. Jane Genzel, the leader, and everyone else made me feel at home and I learned a lot from all those talented people, listening to them read their writings and then reading mine to them.

To my publisher, Vantage Press, who walked me through this first book, my thanks and hope we can do it again.

For those whose names have slipped from memory, I thank you for being there when needed. Catering is not something you do yourself, but with the Lord's help and countless people working together it can be done. Thank you, again.

So You Want to Be a Caterer

Gratuity Included

"Hi, Diane."

It was my friend, Annette, from high school, on the phone.

"A bunch of us from the sorority are getting together at Lu's place for a pot luck next week and we were wondering if you could bring the German potato salad?"

I thought for a moment before answering as I had never made German potato salad. Couldn't be very hard, though, and shouldn't be too expensive—so, "That sounds like fun and sure I'll make the salad. What time are we all going to meet at Lu's place?"

Little did I know that recipe would be the basis for my eventually going into the catering business. Of course there were other factors, but having one good recipe that people always want you to bring or supply is like a shot in the arm. My mother is the one that said to me one day, "One day you are going to make money from that dish." She was right. It got my family through college—along with other recipes also.

You see, I'm a firm believer in the Lord and how He leads our lives if we but let Him. For me, He not only closes one door but opens another to get me going.

My catering was simple when I look back at the weddings and showers that I used to do. I always wanted the bride and groom to have a beautiful reception but because I was aware of the cost of everything I tried to do it as nicely as I could but not charge an arm and a leg in so doing. So my help came from friends, family and students from the local high school. People who wanted to make a little extra money but also knew it was only part time. We took pride in each occasion that we did and were tickled pink when a bride and groom would call afterwards to thank us again. Even now, years later when we happen to meet them again, it is with fond memories of little things that happened.

If this book can help someone who would like to make some extra money from their home I will feel even better.

If you are going to be using your home as the place where you will be doing all the cooking and planning, invite the prospective clients to your home so they can see for themselves what you do and where you do it. Always write out a contract not only to safeguard your client, but also yourself as well. I always required a small deposit ($50.00). This would guarantee them the date but it would not be returned if they cancelled less than two weeks before the date. If I got another client to fill in the time they would get their money back, but if not I kept it. This only happened once as I remember, and it makes for a serious arrangement.

My prices were reasonable and depended on the menu. You will have to make your own guidelines here. It seems to me that weddings that are taking place now are way out of sight cost-wise. It seems that everything is planned so far ahead that they lose sight of what a simple, happy occasion it can be. Even the toasts and memories that are part of the occasion are scripted. What happened to spontaneity? All those little things that are said from the heart, they may not be perfect but they are certainly more fun to remember and laugh about than something off a printed page.

Decorations were made to match the colors of the bridesmaids' dresses and my mother-in-law was a whiz at making decorations out of nylon net for around the candles on each table. I used a heavy paper plate, double thickness napkins, a good plastic cup with handles (no Styrofoam), heavy plastic knives, forks and spoons and a cloth tablecloth. We would all go to decorate the night before, unless it was held in a home, and we would have a lot of fun doing it. The young people who were working for me, as I said earlier, came from our local high school. I approached the school guidance counselor and asked for anyone that was a good student and could use a job. Two of my best young men, David and Jeff, came from that meeting. My kids helped—Vicki, Deb and Bob—and young people from church and three very good cooks, relatives and church friends. All in all there were about twenty people that I would have working, or on call, at any one time. Most of our affairs took place on a Saturday night and I knew from the time that we started serving the food it would be four hours later that we would be emptying the cars and putting things away. Everyone would do the putting away and splitting the leftovers (the families at home really got a kick out of the leftovers) while I sat down and made out all the checks to pay everyone. On Sunday I would sit down after church and see if I had made any money and how much. Everything was paid for so there were no bills to pay—what was left over was my profit. I always collected the balance of the money for the affair at the time we left the party. I had only one problem over the years with a check that bounced, but eventually I received it after several visits to the client's home.

While doing catering I also had a full-time job. My week went something like this: On Monday and Tuesday I would meet with prospective customers in the evening. On Wednesday I tried to do nothing as far as catering was concerned. Thursday, after work, I would do my shopping for the party to be held on the weekend, and on Friday morning my good friend, Grace, would come in and do the preliminary cooking and baking she could do beforehand. I don't know what I would have done without her; there were times I had to because she would have other plans. I learned to be very flexible. Friday nights we would go and do the setting up of tables and decorating; maybe we would go out for some ice cream later if it wasn't too late. Saturday was the party with everyone coming to my home to finish up the cooking and carting and carrying everything over to the hall, church or home where it was to be held. Then my week started all over again. I will say I was much younger then and after my divorce the money was needed to put my kids through college.

Catering

When I first started writing this book I primarily wanted to share my experiences with others who wanted to do the same. Catering from your home can be a challenge, and if my book would help someone think it through before getting involved, it would give me some satisfaction knowing that maybe I had helped someone make a decision.

I do have to tell you that while I was actively involved in my business, a relative of my daughter Vicki, and her husband, invited us to Wisconsin to come and tell them what I did and how I did it. (Reminds me of my Aunt Pearl who was always asking me, "What do you do?" Then when she came to Vicki's wedding she pulled me over to the side as we were doing the catering and now she asked, "How do you do it?") Anyway, Vicki, Al and I went to Wisconsin and after having coffee proceeded to tell them what I did. How I did my marketing, where I purchased my meat and food, how many people worked for me, how my day went when catering, paperwork, just everything I could think of. Well, when I finished the two ladies looked at me without hesitation they both said, almost together "That's too much work." So, I guess you can say I have helped someone make that decision already.

To get back to the reason for the book. As I have progressed in my writing I realized that it could also help brides and grooms and their families make up their minds about how they wanted to go about making their big day memorable.

Is it going to be a big lavish affair that will have everyone talking about it for a long time, or will it be something a little more simple also filled with love, but easier on the pocketbook. Cost was always a big part of the planning as far as I was concerned. There are so many ideas floating out there—all of them beautiful—but what is most important to them both? I have to say with three children and my own wedding many years ago I do have some comments.

First, my wedding was held at a church in the Austin area of Chicago and the reception was at the Norske Club on the Northwest side of Chicago. My father was in the restaurant business and a member of the club. The cost was $2.50 a person and was a lavish buffet of all the Scandinavian goodies. It was held on a Friday night as even back then halls were hard to get, but we didn't know that so we had to settle on that date. There were two hundred people at the reception and I can still hear my friends saying "I hope the buffet stays open after midnight so we can have some meat." Of course they were Catholic and the rule of no meat on Friday was still in effect. There was plenty of other foods: shrimp, cold salmon, herring etc. but that was not the roast beef and ham that they were looking at. The cake had been given to me as a wedding present by my dad's friend who managed Henrices' restaurant downtown. It was baked by their head baker and had spun sugar orchids all over the cake. I, being eighteen and very spoiled, took everything as if it was just a cake; I don't even remember thanking Fred and Hannah that night for it. This was also right after the World War II and everything was still a little crazy; we were all so happy that our fellas were home and did not have to go off to war.

My cousin Bob's wedding was held at Trinity Lutheran Church and the reception was held downstairs. My Uncle Thor, who knew we were having financial troubles, asked me "Don't you wish you had had a wedding like this and saved all that money you spent on that large wedding you had?" I have to admit he hit the nail on the head. So even though it probably helped my dad business-wise to do it the way we did, it would have been nice to have a little of that money at that time.

My daughter Vicki was the first to be married. She chose Bethlehem Lutheran Church for the wedding and the reception. This was handy for everyone, she and Al were going to attend this church after their marriage. My friend, Grace, gave her worktime that day as her gift to the bride and groom—and it was the nicest one they received because it came from the heart. Everyone had a good time and plenty to eat and this is where my Aunt Pearl saw "what I did."

Deborah was the next to get married. She and Bud were going to be living at an apartment complex that had a recreation room and pool. As it was summer they wanted to have the reception there. I seem to remember a little problem with enough tables and chairs so we had to rent some. They were married at our family church, St. Stephen's, in Carpentersville and then we went on to the apartment complex in one of the suburbs. It was a lot more time-consuming and the working space was not very good, but Grace came through again giving her time as the wedding gift. Bud's mother made mint candies for each table in the shape of wedding bells and hearts and in soft colors so the tables were very festive-looking. Because of the logistics it was a much harder wedding to do, but beautiful nonetheless.

Bob was the last to get married and hallelujah it was not my place to do the catering. All I did was give the rehearsal dinner at the hotel we were all staying in. The reception was held at a very nice restaurant and it was kind of fun just being a guest. My Aunt Hazel was a little worried as her husband, my Uncle Thor, had a lady, his age, flirting with him. It was really comical. Aunt Hazel kept sitting closer and closer to him so this lady would know he was taken. This was also the wedding where the Catholic priest who was giving communion came to me in my pew so that my Uncle Leonard, a Lutheran pastor, and Aunt Pearl couldn't see. How things have changed. Now priests and pastors perform the wedding ceremony together and also give Communion together. That is one ritual I am glad to see changed.

So, as I look back at some of the weddings, in fact all of the weddings, it is interesting to see which ones lasted and which ones didn't. My marriage lasted 24½ years so a lot of money spent doesn't guarantee longevity. All my kids are still married as of now. We can't always predict the future. What I can say with confidence is that they all believe in the Lord which is my prayer for them every day. That is what is important, not the size of the wedding or the amount of money spent. Some people are more able to spend a lot and it doesn't make any difference in their lifestyle. But everyone should know that there are many ways to do a wedding and have it as beautiful as you want. It just takes a little time and effort.

Learning the Hard Way

It was 1967. The year of the paper dress. This I had to try as I was searching for the look that I wanted for catering. The company party that I was scheduled to do that week was the perfect time. It wasn't a large party; just the people in the offices. We were serving fried chicken as one of the entrées. This was the one item I did not prepare myself, but contracted with Paul's Deli to make it for me. His chicken was delicious as were a lot of other items he had. In fact, he wanted to go in partnership with me, but he wanted to do it the easy way by buying most of the items from other purveyors. This was not what I wanted to do, so we never could have a meeting of the minds.

Well, I arrived at the office where the party was to be held. I was carrying in things when I felt something rip. My dress! What to do? Well I grabbed my coat and thought, *I'll just tell them I feel cold.* The door opened and the secretary I had been dealing with popped her head in the door and asked, "Aren't you a little early?" I replied, "No, I always get here ahead of time to make sure everything is going on schedule." She hesitated for what I thought was a long time and then said, "But twenty-four hours?" I had the wrong day! I excused myself quickly, got to a phone and cancelled the chicken, called the rental place and explained my problem and they graciously let me keep everything for the extra twenty-four hours without charging me extra.

My face was red for having made the mistake but it taught me a lesson. Always write up a contract with everything listed, not just a receipt for the down payment with notations made on it.

Soon after I picked up some printed contract forms that made it easy to just fill in the correct information. I used them for years, all the rest of the time I was in catering.

The next day the party went off just fine, but I think I lost some credibility about being very businesslike. Oh, yes, no more paper dresses! A skirt and blouse worked out just fine from then on. That's what happens when you try something different and unproved.

Soldier Field

"Mom, there are two gentlemen here to see you."

My daughter, Vicki, had come into the back room where I was writing up some menus to be used in the ice cream store to let me know someone was here to see me. *Probably some salesmen*, I thought, but I got up and went through the swinging doors to see who it could be. Was I ever surprised. These two men were from Detroit, Michigan and were calling on caterers in the area to see about using their services to provide food for a group of actors that were going to be filming a commercial at the Meadowdale Racetrack in Carpentersville, our town. They wanted fifty box lunches for each day of filming. This included the crew that would be doing the filming and the actors. There was one stipulation. We would have to cater one of the days at Soldier Field in Chicago. Wow! Soldier Field! My dad would love to see me be able to do that. I wrote down all the particulars and said I would give my quote the next day.

What excitement! Now to figure out how it could be done. That many box lunches for each day . . . of course I didn't want to repeat any of them . . . my mind was going in circles. Well, let's see, I could use florist boxes to put the lunches in. They would hold a couple of sandwiches, something sweet, and napkins, and any cutlery that would be needed. Also some penny candy and a piece of fruit. Once I got started I kept thinking of all kinds of goodies that could be included. People working would like to have something good and different each day . . . something to look forward to. So, I came up with a menu for each day and what it would cost and had my proposal ready. The gentlemen (isn't it funny I can't remember their names?) looked it over and after conferring a few minutes said that they would like to have me do the job. Glory, glory! I was on cloud nine. A deposit would be in the mail the next day along with the contract. Anyway, I was on my way—now to make sure it all went well.

I needn't have worried. All the comments when handing out the boxes were good. Several actors said, "This is enough for two meals, lunch and supper." I imagine some of them were on tight budgets. When the day came to go to Soldier Field there were reporters along, and they were asking if they could buy some box lunches for themselves. This is where I missed the boat because it was so time-consuming to get the original order together I never thought about making extras. My loss!

The menu varied each day. At least two sandwiches, condiments, celery, carrots, pickles, olives, bags of potato chips, cookies, cake or cupcakes, fresh fruit, napkins, cutlery and always several pieces of penny candy. One day I served fried chicken instead of sandwiches . . . that was also greeted with delight. It was fun to see them all enjoying everything so much, and I wonder to this day if there were any aspiring actors that became famous later on.

I found out, later, that there had been some questions of having an outside caterer come into Soldier Field, but it was kind of funny to see the powers that be watch me drive up in my old station wagon that was practically falling apart and hand out lunches with two teenagers helping me. At this point we were all wearing granny dresses as our uniforms, so to speak, it must have been quite a sight. I don't think the scene has been duplicated since, but being on the field itself, well I just wish my dad had been there.

The commercial had to do with the Road Runner product and for the final day of shooting we were invited to come to the wrap party that was being catered by the same caterer that did all of LBJ's parties down in Texas. It was eye-opening to see them do things pretty much the same way I did, but of course I'm sure their bill was a lot larger than mine. The food was good and I enjoyed the experience of saying that I had eaten the same food our president ate at his parties.

Catering can lead to a lot of fun experiences and also one or two that are not. But more of that in another chapter.

Recipes

There were several menus that were presented to prospective clients. It, of course, depended on what kind of affair was to be served. The most popular was the buffet with roast beef, baked ham, two kinds of potato salad, cole slaw, fresh fruit bowl or fresh salad with celery seed dressing, rice pudding, home made cookies, or if the family was Italian, a selection of Italian cookies from a very good Italian bakery. Our cookies were made with butter and the favorites were Spritz, Toffee Bars, Brownies, Thumb Print with strawberry jam, Mexican Wedding Cakes, and Miniature Cream Puffs.

All our baked goods were made by my best friends, Grace and Elaine. They had a way with baking that I couldn't duplicate. Our wedding cakes were made by the White Cross Bakery in Dundee. Bill, the owner, would also make specialty cakes for anniversaries of churches. He would go and take pictures of the church and then duplicate it into a very edible copy. He never overcharged for this so I could keep the prices as low as possible. He also made Petit Fours that were not only beautiful but delicious. So many times these little cakes will look good but taste like air when you bite into them. His advertisement for his delicious French doughnuts was "recommended by Dunkin Hands."

When we did small house parties such as showers, or even funerals, we would suggest Chicken Tetrazzini or Chicken à lá King, along with our celery seed dressing salad or fresh fruit and tomato aspic and a selection of cookies and cake. The cookies could always lend themselves to almost any menu and went over very well. We have even done open-face Danish sandwiches which are very good, each sandwich decorated with the condiments that go with it.

When doing an hors d'oeuvre party the selections were Crab Salad Puffs, Ham Salad Squares, Shrimp Salad Triangles, Chicken Salad Rounds, Date and Nut Triangles, Egg Salad Flower Pots, Fannie May miniature chocolates, and a deluxe assortment of nuts, including cashews. If hot hors d'oeuvres were the menu it would include Swedish Meat Balls, Barbecued Ribs, Flaming Pineapple, miniature Meat Ball with sweet onion rolls, miniature Roast Beef and Horseradish sandwiches, and miniature Ham sandwiches with mustard. I depended on the different breads and shapes to make the trays look appetizing and also my three Lundgren girls to make everything look so good. No matter where we were or what we were doing those three girls made everything look like a picture. I would just turn them loose and they would work their magic.

So, you can see I depended on a lot of people to get everything done. I learned from them so if I had to step in when they were gone, I could. My friend, Elsie Lee (who died far too young), worked with me on planning the hors d'oeuvres menu and sharing her chicken salad recipe with me. She didn't like mayonnaise so her secret was using sour cream and almonds . . . delicious. When sharing those recipes with you I will also name the person who gave them to me.

The Lord has been very good to me, letting me be able to do this to get my kids through school and married before giving it up. I could never have done it if I hadn't put my trust in Him. He always sent me someone when I needed them . . . and were least expected.

Chicken Salad—Elsie Lees' Recipe

1 can Swansons boned chicken
1 good T slivered almonds—chopped for crispness
1 small jar pimento—about half
Sour cream to moisten
Salt and pepper to taste

Mix all ingredients so that the mixture will be easy to spread.

This can be used to fill sautéed mushrooms or for finger sandwich rounds, which is the way we used it for the most part.

Take a thin slice of white bread (we used Pepperidge Farm sliced white bread); and also whole wheat and oatmeal for different colors and textures.

One slice of bread will give you the top and bottom for the finger sandwich. I used one sized glass for the bottom and another for the top.

Always butter the bread so the salad doesn't soak into it and it will stay fresh. If you want to work with frozen bread you can use melted butter.

Spread the salad on the large round and add the smaller round on top.

This will take about two bites to consume, delicious, so you will need at least two for each person along with the other hors d'oeuvres you have. This is also good for fancy sandwiches, cut in half or quartered.

Tetrazzini—Chicken or Turkey

1/4 cup flour
1/4 cup butter
1 cup chicken broth
1 cup whipping cream
2 T sherry
2 cups cubed chicken or turkey (can be canned or fresh)
1 can (3 oz.) mushrooms
1/2 cup Parmesan cheese
1/2 t salt
1 lb. spaghetti

Cook spaghetti according to package directions.
Combine chicken broth, whipping cream and sherry.
Add to butter and flour mixture to make the cream sauce.
Put cooked spaghetti into casserole, add chicken, mushrooms, Parmesan and salt. Mix well.
Pour cream sauce over the spaghetti mixture.
Bake, uncovered, in 325 degree oven until bubbly.
Dot with butter—use bread crumbs for topping if you want.
This is a very satisfying dish. It not only tastes good, but looks good. Combined with a salad, bakery rolls and butter, dessert of cookies, cake or a combination of both.

Celery Seed Dressing

1/4 cup celery seed
1 1/2 cups oil (vegetable)
1/2 cup vinegar
3 t dry mustard
1 finely chopped onion (I use yellow onion medium size)
6 T sugar

Mix all ingredients together in a blender or with an egg beater.

Pour over mixture of lettuce, mandarin oranges, French onions and pecans. If you don't need all of the dressing it will keep in the refrigerator to be used at another time. I use canned French onions, canned mandarin oranges and packaged pecan halves. This was known as Salad "à lá Mrs. K."

Toffee Bars—Grace Dorpal's Recipe

1 cup butter
1 cup brown sugar
1 cup flour
1 egg yolk
1 t vanilla
6 regular Hershey bars
3/4 cup crushed pecans

Cream butter and sugar.

Add egg yolk, flour and vanilla. Beat until smooth and spread on cookie sheet or a 13 x 9 x 2 inch pan. Bake at 350 degrees for 20–25 minutes (until lightly browned).

While still hot cover with six regular size Hershey bars and when they have melted spread evenly over the top. Sprinkle with crushed pecans and cut into squares while still warm. You can cut them the size that you need for the number of servings you want. If for a cocktail party the size would be smaller than if you were serving a buffet.

I have to say these are one of my favorites, and if my friend Grace didn't live all the way in Georgia I would love to have her make some for me.

Champagne Punch

1 can frozen orange juice made up
1 can frozen lemonade made up
1 cup Countreau
1 bottle champagne (we used André)
1 quart white wine—Chablis

Mix all ingredients in a large punch bowl. Fill with ice and the punch is ready.

I have to explain about serving liquor. I did not have a license so what I would do is order the amount needed and present the bill to the customer. There was no added charge for going to the liquor store and making the arrangement. My help was already at the party so putting the punch together was no problem, and anything left over would stay with the customer. But this is so good there really was no problem there. I had one customer who wanted to serve beer along with the champagne punch. I was very adamant that this should not be done, but a few days later I read where this had been done at a well known function. I immediately called, apologized and told them I could let them order the beer to be served as they knew what brand, or brands that they wanted. It was a very nice affair held at their home and I also did other parties for them.

The Church Lady

This chapter not only deals with the good-hearted ladies of all the churches that we catered in, but also to let anyone thinking about catering know what a good place a church is to hold a wedding and a reception.

In planning a wedding a lot of things have to be taken into consideration. The couple should not only think about themselves, but their guests as well. Is the church and reception area easily accessible to all ages? If not, don't feel bad if someone you would really want to come to your wedding would have to refuse because of some reason such as health or disability. You can be sure they would be there if there was any way for them to attend. That doesn't mean that you should change your plans because this is the way you have always dreamed it would be, but to try and understand the reason someone has had to send their regrets. Let them know you will miss them and to please remember them in their prayers on the day of the wedding.

The church, if you should decide to have your reception there, will make you feel more than welcome 90 percent of the time. They undoubtedly have a group of ladies that will put on a lunch after a funeral, dinners for special occasions and I'm sure Easter breakfast. So, the majority of the items you need to serve with will already be there. This means you have to be very careful not to misuse this privilege. When we would go into a church for the first time we always expected two or three church ladies sitting in the kitchen to watch what we did and what we used. They would sit there with their cup of coffee (we would see to it that they had something to go along with the coffee) and just watch and listen. The next time we would come there might be one or two ladies sitting and watching and when we came the third time . . . no more church ladies. They had gotten to know us and trusted us with everything in their kitchen. Now I know it takes a long time to get a kitchen workable for everyone to be able to use it, so we fully understood their concern. In showing this concern you could be sure that the kitchen was a very easy one to work in. The reception area would also be well thought out. Not only were the bride and groom able to have a good time and show their guests a good time as well, but the church was gaining as well with the nominal cost to use everything.

Happenings

"Deb, did you see that?"

My daughter and I were manning the ice cream store for the afternoon. It was during a slow time and I had just looked up at our plate glass window that had "Mr. K's Old-Fashion Ice Cream" in bright letters. We were close to the junior high school so it was a place where kids would stop after school. We only had a few stools to sit on. The décor was almost like a bar with two six-foot mirrors behind the bar. It was an inviting place and I enjoyed taking customers' orders and seeing them enjoy the good ice cream. It was too bad it didn't make any money. It took only eighteen months to lose $5000.00. But we did have fun with it and it led to the catering. Even now, as I look back, if I had insisted on having soft serve ice cream along with the regular Sealtest ice cream it might have been successful. Oh, well, that isn't what I started to tell you about.

Deb looked up from what she was doing in time to see the two girls laughing after they had thrown their McDonald's shakes at the window and take off running away towards the field that was at the end of our buildings.

Well, we both looked at each other and without a word both hiked up our skirts (we were wearing granny dresses at the time) and we both took off after them. It surprised me that we were able to catch them about a half a block into the open field and bring them back to the store. They were going to clean up the mess they had made! We were adamant about that . . . and to my surprise they really didn't give us much argument. We got them the materials needed to wash the windows and pretty soon they were done—even with some kidding from others who were passing by or came into the store to get an ice cream. One of us stood watch outside, but it went pretty well considering.

The store closed with many tears by me, but it was not meant to be. The catering took off and became a business that made a little money. One day I got a call from someone I had to share with Deb. "Guess who called to have us cater her wedding . . . you guessed it, one of the girls who had thrown the McDonald's shake at our window. I don't think she thought I would remember her, but I did." The wedding went off without a hitch and both she and her new husband couldn't thank us enough. Kind of funny when you think about it.

The Baker Hotel

I had received a call from a mother and daughter who lived in St. Charles, Illinois. This was a little out of my way but I thought it might be something interesting and who knows where it could lead. They had mentioned the Baker Hotel as the place they wanted to hold the wedding reception. I was very curious to see what the hotel looked like from the inside as I had only gone by it a few times. From the outside it reminded me of the old Edgewater Beach Hotel in Chicago, which was on Lake Michigan. This hotel was on the Fox River and was much smaller, but I was still curious.

    I met with the clients in their home. It was in a well established part of town and they made me feel very comfortable sitting in the very homey feeling living room. It was just the mother and daughter meeting with me. They explained that the hotel was in transition from being a regular hotel to being a upscale retirement hotel and all the plans for food had not been decided as yet, so that we could come in and rent it for the wedding and use all the kitchen facilities, which the wedding party would pay for.

    They knew just what they wanted. There was a nice lobby, they told me, where they would greet the guests as they arrived. When everyone was there they would then enter the ballroom, where all the tables had been set up. They would ask everyone to be seated and then just before the wedding party entered they wanted me to turn on the lights that were in the glass brick floor so that they could walk in with a blaze of color.

    Well, this was something new. I had never heard about the

beautiful floor, so as we drove over to see everything my head was just about bursting with ideas. We were greeted by the person in charge and she again explained that this was just a temporary option for the hotel until a regular caterer could be agreed upon. The inside was just like a miniature Edgewater Beach Hotel. It dated from the 1920s and was almost like a movie set, bringing back memories of all the old black and white movies of that era. The place was spotless. The basement, where all the lights were controlled, couldn't have been cleaner. I couldn't find anything that even looked like dust in the whole place. Someone was very proud of how it had been kept. The kitchen was just as clean; it was a little old fashioned, but workable. The dining room, or the ballroom, was oval with a balcony all around with tables set on it overlooking the river. There were also tables beneath the balcony itself, all with the same view.

We decided on the date and the menu, and after conferring with the person in charge I felt eager to get going on the rest of the plans.

The day arrived and my crew arrived with me a few hours before so that we could get the lay of the land, so to speak, and to make sure that I knew how to work the lights in the floor. We set the buffet table on the far end of the floor so that it didn't interfere too much with the bridal parties' table and most of all with their entrance.

The time came and as planned the party started to enter. I was downstairs and waited for my cue to let me know when to turn on the lights. After the bridal party had gotten to their places at the doorway leading into the room I heard Vicki say, "Now." *Please let these work,* I prayed, as I pulled the switches. I needn't have worried. All of a sudden I heard the loud "oohs and aahs" as the bridal couple walked into the room. It had all been worth it. The tables with white tablecloths, the candles in different colors and the multicolored lights in the floor were a sight to behold. I don't think anyone at that wedding will ever forget the special entrance of that bride and groom.

I've always meant to go back to the hotel and experience the dining room as it now is. It is a well thought of retirement hotel, and it is open to the public for lunch and dinner. As I live in a retirement community I can see the advantage of having other people come in who want to enjoy a good time and mingle with the retirees. It is a little on the expensive side, I have heard, but well worth it if you have the money to live as if you were back in the 1920s with all the memories that go with it.

The Table

I have mentioned before that I depended on a lot of good people to work with me. Some of them still like to kid me about the way I would do things.

For instance, my son-in-law, Al, and his kids, Lisa and Lora along with my daughter Vicki, would help me set up all the tables at the church or hall where we would be serving the next day. He would kiddingly take bets as to how I would want the tables set up: on an angle, or straight. He insists that no matter which way he chose I would want them the other way. He is a good cook and if times had been different I'm sure that he would have been a good manager of a very nice restaurant.

My sister and brother-in-law would help me occasionally, but they really didn't want to do it, so it was only rarely I could get them to work with me.

There were two boys that I counted on most of all, David Christensen and Jeff Pepper. Sadly I have lost track of them, but if they continued the work ethic that they showed when they worked for me, they have done okay.

On Saturday mornings we would all sit around the "temporary round picnic table" that was in our kitchen, peeling potatoes or making fancy sandwiches. I say temporary because my husband had sold a lot of cars the year we moved into our house and instead of spending money on a kitchen table the prizes that year were for outside furniture. So into our kitchen it went and stayed there for eighteen years until my daughter Deb took it and finally used it as it was meant to be used, as a picnic table. We would all sit around laughing and talking while working. Grace, Elaine, Marianne (another lady from church) and anyone else that was there to work. The time went quickly as everyone knew what their job was. Grace had been there the day before and had done what baking was to be done, Elaine had done her baking at her house across the street. Grace would get all the bowls ready that we were going to use and the platters. I would slice the meat that I had baked during the night and the boys would start loading the car.

My job was to make the German potato salad. Now I know Grace or Elaine could have made it, but I was very positive that I wanted to know that I had made it. Silly, I'm sure, but that was the way it was.

One evening another friend of ours from church was going through a hard time. She was an actress through and through but we were concerned about her because a one time boyfriend of hers was getting married to someone else that night. So we thought, why not have her work for us for the night to see how she did. Well, I asked her to stand by the buffet table and to let us know when dishes had to be refilled or replaced. She took to it like the actress she was. She stood at the head of the table and as each person came up she nodded to them with a beautiful smile to let them know which side of the table they should approach. That was it, smile and nod, smile and nod, while we ran around trying to keep the table filled.

There was a time we were doing a home party for someone who was a backer of the Dundee Scots, our high school band. They were going to play that night at the home (not the whole band but a good part of it). Anyway, everything was going just fine and on schedule when the lady of the house came down and asked us to hold off on dessert because the band was going to be late. We were serving Chocolate Mousse and it really had to be just at the right temperature. Well, we put it back and a little while later the gentleman of the house came down and said it was OK to serve. We weren't going to wait for the band. Out came the Mousse when the lady of the house came down and told us we had to wait again because the band was just pulling up. Back in the refrigerator it went one more time. At this time, Grace, who is a perfectionist especially as far as her food is concerned, was getting a little upset. She told me the next time they say serve, we will serve, no matter what. Which is what we did. It was a fitting ending to the meal—while listening to the music they ate their Chocolate Mousse. We got to listen, but no Mousse.

So, lots of things can go wrong, especially if you serve on New Year's Eve at a tennis club where the games that were part of the evening fun lasted longer than they were supposed to. The plan was to have two different serving times. While one group was playing a game of tennis the other group would be having the buffet. That would have been fine, but a little too much celebrating with the liquid refreshments (which we had no control over) and everything went to pot. The food was good. My son-in-law was carving the rolled roast filled with a homemade stuffing and was losing his cool as people would say it was lamb and he would say "no, it is roast beef" and then little arguments would start as to who was right. We laughed it off realizing the condition most of them were in at this time of the night. This is the only time my daughter-in-law-to-be worked with us—and needless to say she made a very good insurance agent. One night was enough for her.

My son-in-law, Bud, also tried for one night and that was enough for him. It was a very large church affair and I had told the committee when making arrangements that I could not arrange for water at each person's place, but that I would set up a table with water and glasses. Well, as I do when I'm in a hurry, I didn't make myself clear to Bud and he started to fill a glass at each place. He wasn't too pleased when I told him he would have to remove them. So one other in-law quickly lost interest. His wife, my daughter Deb, really wasn't into it either so they have been very successful in their chosen fields. She as a high school math teacher, he as a man who travels, even to China, for his company all the time. This year he brought me authentic Chinese chopsticks and holder which I didn't have. I love Chinese food and you would think I would have had my own chopsticks before this as I worked for a gourmet Chinese restaurant as a bookkeeper and caterer for over four years. That is another story.

Not everyone is suited to this kind of work. My fault was in not appreciating enough what I did in catering, and to charge more for the little added things that I provided. But I will never forget the fun part of the work and meeting people years later who would come up to me with stories that I had forgotten.